The animals were excited —
they simply couldn't rest.
Soon the contest would begin,
to see whose jump was best.

Pete the Puppy was the first
to come up to the line.
But his jump was a short one,
and he gave a little whine.

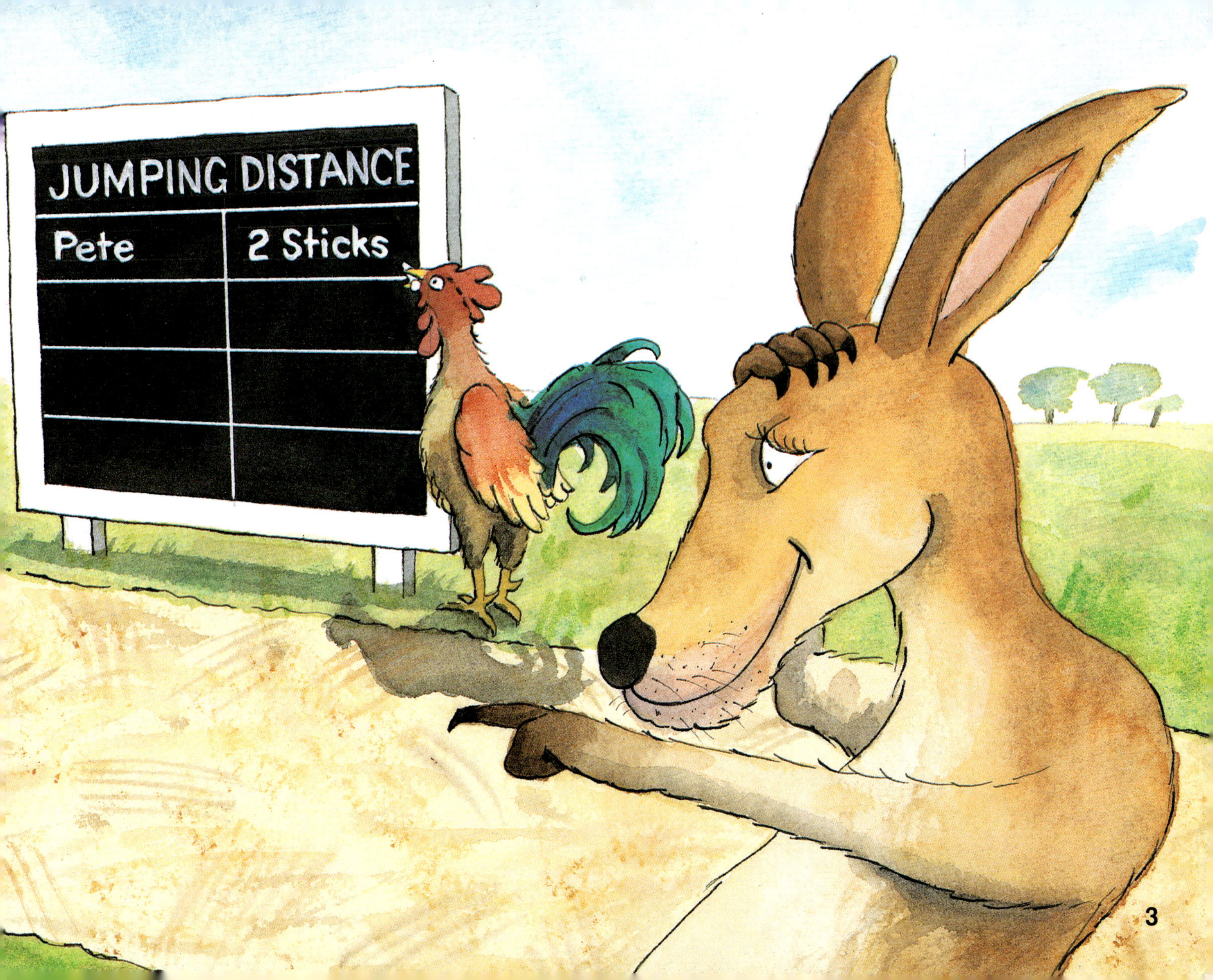
JUMPING DISTANCE
Pete
2 Sticks

Rosco Rooster was the next
 to line up for a jump.
He flapped his wings —
 then off he went —
  and landed with a bump!

JUMPING DIST	
Pete	2 St
Rosco	4 St

5

Katy Kid was next —
   she had lots and lots of spring!
To measure her long jump,
   they used a measuring string!

JUMPING DISTANCE	
Pete	2 Sticks
Rosco	4 Sticks
Katy	6 Sticks

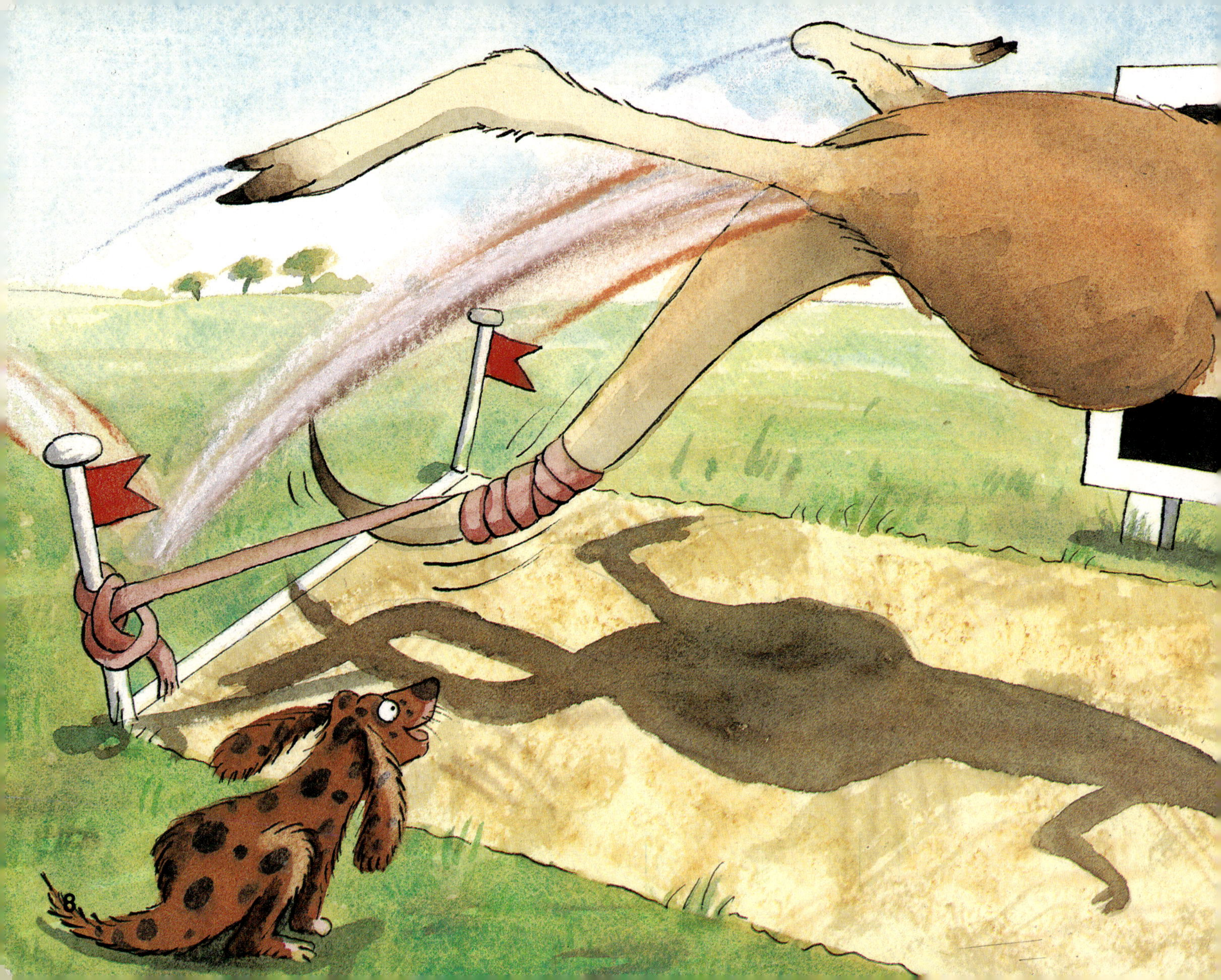